SCIENTIFIC AMERICAN | EDUCATIONAL PUBLISHING

SCIENTIFIC AMERICAN INVESTIGATES CAREERS IN SCIENCE

ASTRONOMER

BY ROSIE BANKS

Published in 2025 by The Rosen Publishing Group
in association with Scientific American Educational Publishing
2544 Clinton Street, Buffalo NY 14224

Library of Congress Cataloging-in-Publication Data
Names: Banks, Rosie, 1978- author.
Title: Astronomer / Rosie Banks.
Description: New York : Scientific American Educational Publishing, an imprint of Rosen Publishing, [2025] | Series: Scientific American investigates careers in science | Includes bibliographical references and index.
Identifiers: LCCN 2024036204 (print) | LCCN 2024036205 (ebook) | ISBN 9781725352322 (library binding) | ISBN 9781725352315 (paperback) | ISBN 9781725352339 (ebook)
Subjects: LCSH: Astronomy–Vocational guidance–Juvenile literature. | Astronomers–Juvenile literature. | Astronomy–Juvenile literature.
Classification: LCC QB51.5 .B36 2025 (print) | LCC QB51.5 (ebook) | DDC 520.23–dc23/eng20240814
LC record available at https://lccn.loc.gov/2024036204
LC ebook record available at https://lccn.loc.gov/2024036205

Designer: Leslie Taylor
Editor: Therese Shea

Portions of this work were originally authored by Nicole Shea and published as *Be an Astronomer*. All new material in this edition is authored by Rosie Banks.

Photo credits: Cover (main) AstroStar/Shutterstock.com; series art (background) jijomathaidesigners/Shutterstock.com; p. 5 Stock-Asso/Shutterstock.com; p. 7 Arina Gladyisheva/Shutterstock.com; p. 8 NASA/File:James Webb Space Telescope 2009 top.jpg_commons.wikimedia.org; p. 9 rinaK/Shutterstock.com; p. 10 Nazarii_Neshcherenskyi/Shutterstock.com; p. 11 NASA/File:Southern Ring Nebula (NIRCam Image).png_commons.wikimedia.org; p. 13 (sky) Skylines/Shutterstock.com, (inset) AstroStar/Shutterstock.com; p. 15 VikiVector/Shutterstock.com; p. 16 buradaki/Shutterstock.com; p. 17 (main) Credit: NASA/JPL-Caltech/SSI/File:8423 20181 1saturn2016.jpg_commons.wikimedia.org, (Saturn with Titan) Cassini Imaging Team, Image Credit: NASA/JPL-Caltech/Space Science Institute/File:Titan in front of the ring and Saturn.jpg_commons.wikimedia.org; p. 19 (asteroid) NASA/JPL/File:243 ida crop.jpg_commons.wikimedia.org, (comet) E. Kolmhofer, H. Raab; Johannes-Kepler-Observatory, Linz, Austria/ File:Comet Hale-Bopp 1995O1.jpg_commons.wikimedia.org, (meteor) NASA/ESA/ File:Meteoroid track through aerogel from EURECA mission.jpg_commons.wikimedia.org; p. 21 ManuMata/Shutterstock.com; p. 22 muratart/Shutterstock.com; p. 23 Mariia Domnikova/iStockphoto.com; p. 25 (main) NASA/CXC/MIT/F.K.Baganoff et al./File:The supermassive black hole at the center of our Galaxy. (2941518598).jpg_commons.wikimedia.org, (inset) EHT Collaboration/File:Primera imagen de Sgr A* tomada por el Event Horizon Telescope Collaboration.jpg_commons.wikimedia.org; p. 26 Moshe EINHORN/Shutterstock.com; p. 27 MarkoBeg/Shutterstock.com; p. 28 Artem Oleshko/Shutterstock.com; p. 29 Purdue9394/iStockphoto.com.

Printed in the United States of America

CPSIA compliance information: Batch #CWSA25. For Further Information contact Rosen Publishing at 1-800-237-9932.

CONTENTS

Words in the glossary appear in **bold** type the first time they are used in the text.

AN EXCITING CAREER

Astronomy is the study of the universe outside Earth's atmosphere. That includes the things we can see and the things we can't. Scientists who study the universe and the objects in it are called astronomers. Astronomy is one of the most exciting science fields. New **technology** is making it possible for astronomers to make discoveries often.

Whatever you're wondering about the universe, astronomers are studying it. Do you wonder about life on other planets? So do astronomers. How about giant space rocks hitting Earth? Astronomers are on it! If you're interested in unlocking the mysteries of space, you might want to be an astronomer.

FUN FACT

IN 2020, 12-YEAR-OLD RAFAŁ BIROS BECAME THE YOUNGEST PERSON EVER TO DISCOVER A **COMET**.

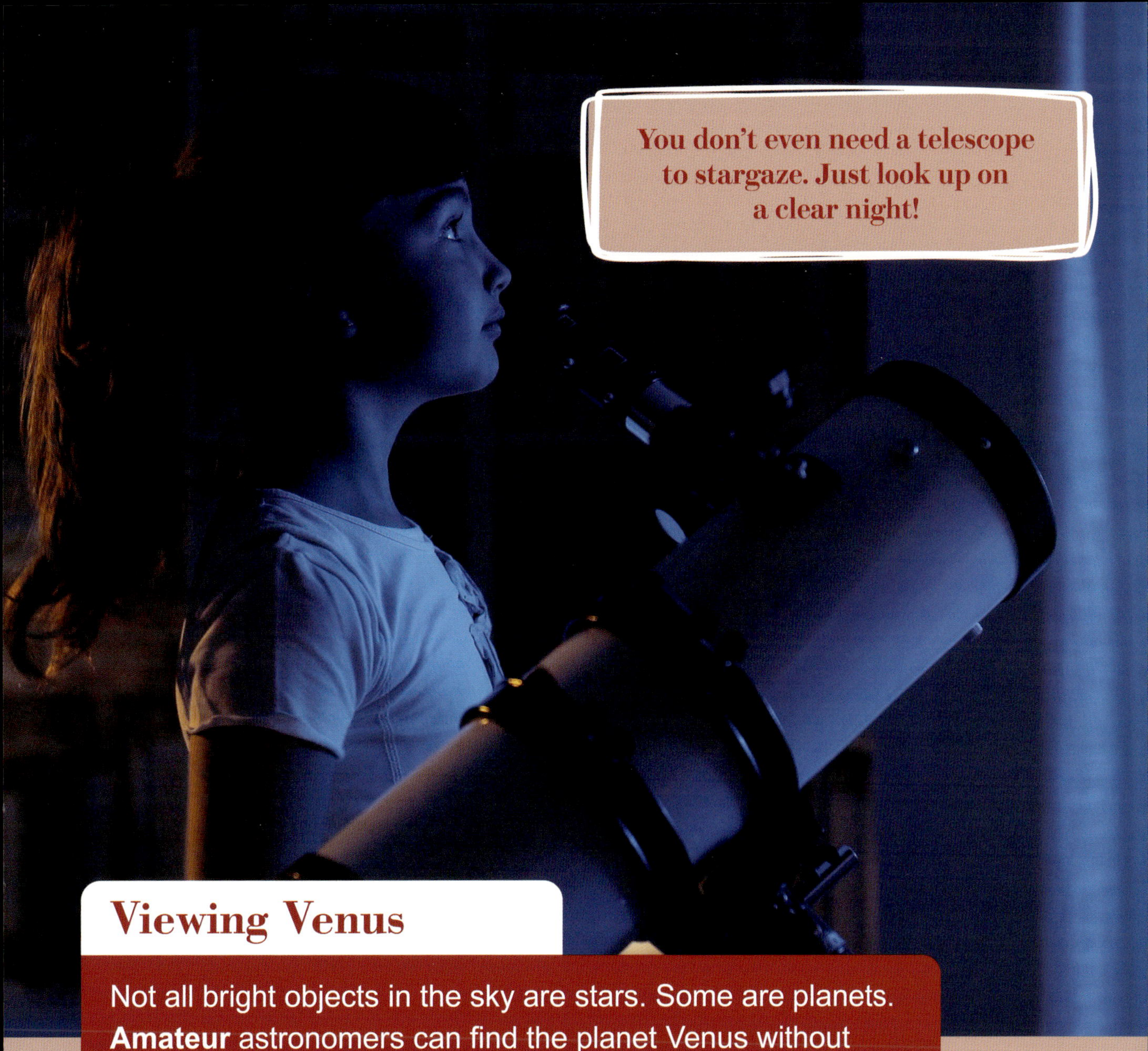

You don't even need a telescope to stargaze. Just look up on a clear night!

Viewing Venus

Not all bright objects in the sky are stars. Some are planets. **Amateur** astronomers can find the planet Venus without a telescope. It's often the first point of light you see when it starts getting dark at night or begins to grow light in the morning. The planet reflects the sun's light.

EARLY ASTRONOMY

Even before people invented systems of writing, they studied the stars. Some saw patterns that reminded them of animals or story characters. A named pattern of stars is called a constellation. Constellations are helpful for locating stars in the night sky. For example, Polaris, or the North Star, is in the constellation Ursa Minor.

As stargazers watched how most stars seemed to move from night to night, they noticed the star called Polaris is fixed in the night sky. Travelers could walk toward Polaris and be sure they were headed north. People also discovered stars helped them keep track of time, which was helpful for farming and **navigation**.

CONSTELLATIONS OF THE NORTHERN HEMISPHERE

Cassiopeia, Cepheus, Draco, Ursa Major, and Ursa Minor are constellations that people in the Northern Hemisphere, or the northern half of Earth, can see all year long.

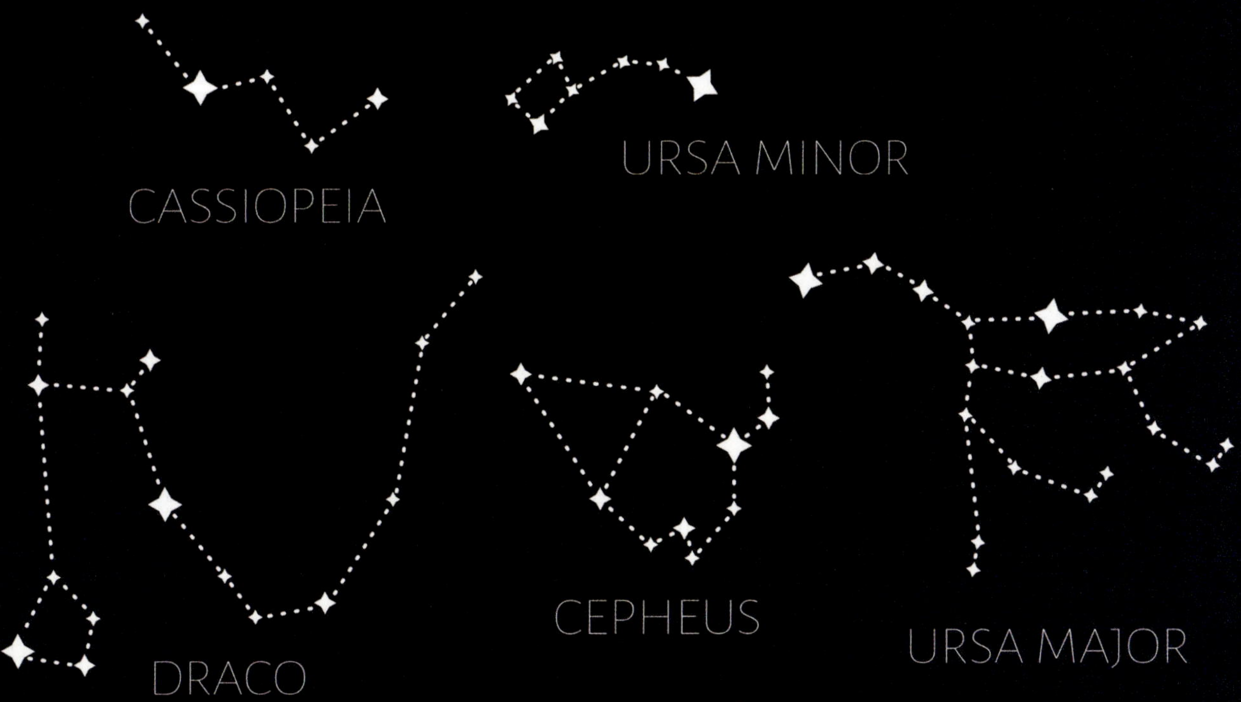

FUN FACT

URSA MAJOR, WHICH MEANS "GREAT BEAR," IS SOMETIMES CALLED THE BIG DIPPER BECAUSE SOME OF IT LOOKS A BIT LIKE A SPOON. BUT THE SPOON IS JUST PART OF THE WHOLE CONSTELLATION. URSA MINOR IS NEAR IT.

TELESCOPE TECHNOLOGY

Early astronomers could only use their eyes to make observations. In the early 1600s, people began making and improving telescopes. In 1610, an Italian scientist named Galileo Galilei used a telescope he had built to discover four small bright moons orbiting, or circling, Jupiter.

Galileo used **optical** telescopes as many astronomers do today. However, these aren't helpful during the day or on cloudy nights. Other telescopes, including radio telescopes, can be used in daytime and in bad weather. Radio telescopes detect radio waves, a kind of invisible light wave, that come from objects in space. Astronomers can build an image of space objects using these radio waves.

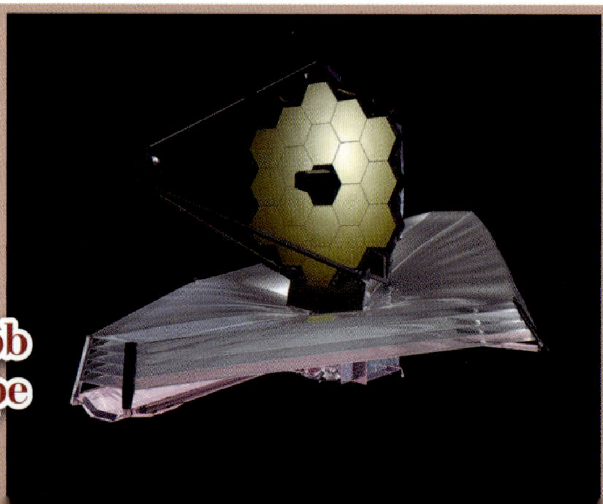

James Webb Space Telescope

The most powerful radio telescope on Earth is the Very Large Array (VLA) located in New Mexico. It's made up of 27 antennas, several of which are shown here.

The JWST

Earth's atmosphere can make it hard for some kinds of light waves to reach the surface, where ground telescopes can detect them. That's why some telescopes are placed in space. The James Webb Space Telescope (JWST) orbits the sun, about 1 million miles (1.6 million km) away from Earth. It mostly observes light waves called infrared waves.

HOW FAR?

Powerful telescopes such as the JWST are needed to see objects light-years away. A light-year is a measurement astronomers use to record an object's distance. It's the distance light travels in 1 year, which is about 5.88 trillion miles (9.5 trillion km). The farther away something is, the longer it takes for its light to travel to us. When astronomers observe light from a star four light-years away, they're seeing what it looked like four years ago.

The distance from our sun to the nearest star, Proxima Centauri, is about 4.24 light-years. The nearest **galaxy**, the Canis Major Dwarf Galaxy, is about 25,000 light-years from the sun.

FUN FACT

ASTRONOMERS THINK THE UNIVERSE IS EXPANDING. THAT MEANS ALL POINTS ARE GETTING FARTHER APART!

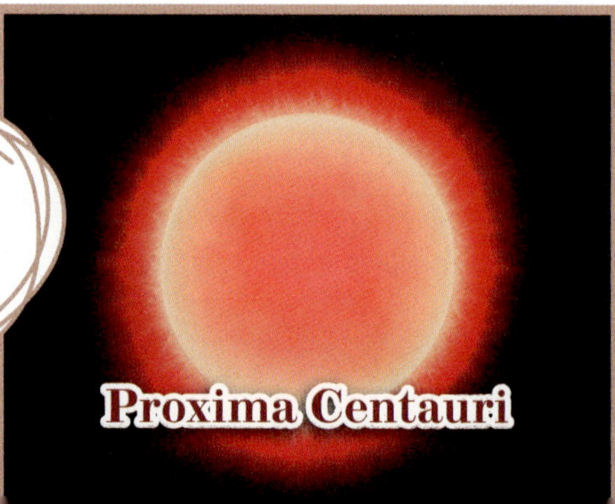
Proxima Centauri

The JWST captured this image of a planetary nebula, an area of gas and dust formed from a dying star. It's called the Southern Ring Nebula. It's 2,000 light-years away!

TODAY'S ASTRONOMERS

The universe is huge! Astronomers usually focus, or center, their studies on a small area. Some focus on planets, others on stars, and still others on the search for life beyond Earth. Some astronomers focus on observing and gathering data, or facts and figures. Others focus on trying to figure out why the universe acts as it does. They use data and math to develop theories, or explanations based on facts.

Today's astronomers spend less time at telescopes. Computers and cameras attached to some telescopes can capture images. In fact, much of an astronomer's day is spent in an office looking at pictures and using computer programs.

Types of Astronomy

Astronomers have knowledge of many kinds of astronomy, but often focus their studies on one of these.

Type	The study of
astrometry	the measurements of positions and movements of space objects
planetary astronomy	planets, moons, and other objects in systems
astrophysics	how the laws of **physics** applies to space objects
astrochemistry	atoms and **molecules** in space
astrobiology	the possibility of life beyond Earth
astrogeology	the **geology** of planets and other space objects
stellar astronomy	the life cycles and structures of the sun and stars
galactic astronomy	our galaxy, the Milky Way
extragalactic astronomy	galaxies outside our own
cosmology	the origin and nature of the universe

When astronomers need to use powerful telescopes, they go to an observatory, a building that houses a telescope. Optical telescopes like this one observe light waves that are visible to the human eye.

PLANETARY RESEARCH

Planetary astronomers study planets, their **satellites**, and smaller objects in planetary systems. This **research** includes how these objects first formed and their current features and movements.

You probably know quite a bit about the eight planets in our solar system. Mercury, Venus, Earth, and Mars are rocky planets. Jupiter, Saturn, Uranus, and Neptune are gas planets. Planetary astronomers also study asteroids, which are smaller rocky bodies that orbit the sun. They may observe asteroids' movements to figure out if their paths are a danger to Earth.

Some planetary astronomers also study planets beyond our solar system, called exoplanets. Scientists have found more than 5,500 exoplanets orbiting faraway stars!

A planetary system is one in which planets orbit stars. Our planetary system, shown here, is called the solar system.

The Bigger Picture

Our solar system is located in the Milky Way galaxy. The sun orbits the center of the Milky Way. There are between 100 billion and 400 billion stars in our galaxy. Some galaxies are much larger, with trillions of stars. Astronomers think there may be between 100 and 200 billion galaxies in the universe!

Planetary astronomers may study moons, or naturally formed bodies that orbit planets. Moons are also called planetary satellites. Astronomers have counted 293 moons in our solar system so far. They've also located more than 470 satellites that orbit dwarf planets, asteroids, and other smaller space objects.

Not all moons are like our moon. Titan, one of Saturn's 146 moons, has lakes. They contain liquid methane and ethane, which on Earth are colorless gases made up of carbon and hydrogen atoms. But Titan also has an ocean of water under its icy surface. Water is needed for life, but so far, there's no signs of life—as we know it.

Pluto (dwarf planet)

FUN FACT

A DWARF PLANET IS A LARGE OBJECT THAT ORBITS THE SUN, LIKE A PLANET DOES, AND HAS MASS AND GRAVITY THAT CAUSE IT TO BE ROUND, LIKE PLANETS HAVE. HOWEVER, A DWARF PLANET CANNOT CLEAR ITS ORBITAL PATH OF OTHER OBJECTS AS A PLANET CAN.

The Cassini **space probe** sent back these images of Saturn, its rings, and its largest moon, Titan. The rings are made of dust, rocks, and ice pulled from Saturn's moons and passing objects such as asteroids and comets.

SORTING SPACE ROCKS

Some people make wishes when they spot a "shooting star." A planetary astronomer would tell them that shooting stars aren't stars! They're meteors, small burning rocks traveling through Earth's atmosphere. If they reach the ground, they're called meteorites.

Comets are large pieces of rock, ice, and other matter. As they near the sun, the ice melts and releases, or lets go of, gas and dust, creating a "tail." Some comets pass by Earth regularly.

Asteroids are made of rock and sometimes metal, mostly nickel and iron. They can be small or large. Scientists think a large asteroid crashed to Earth 65 million years ago, wiping out the dinosaurs.

As of July 2024, astronomers have counted 1,372,879 asteroids in our solar system and 3,949 comets and comet pieces. About 25 million meteors enter Earth's atmosphere every day!

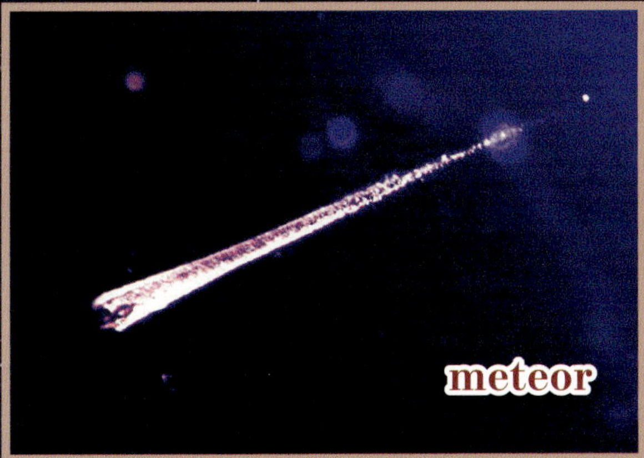

FUN FACT

HOW CAN YOU TELL THE DIFFERENCE BETWEEN A LARGE ASTEROID AND A DWARF PLANET? DWARF PLANETS, SUCH AS PLUTO AND CERES, HAVE ENOUGH MASS AND GRAVITY TO GIVE THEM A ROUND SHAPE. ASTEROIDS DON'T.

STELLAR RESEARCH

Stellar astronomy, the study of stars, includes what they're made of, how they move, and how they die. Our sun is like other stars in that it's a ball of gas that changes hydrogen into helium through a process called nuclear fusion. Energy—light and heat—is a product of this process. By comparing our sun to many other stars, astronomers have discovered our sun is a medium-sized star. Some stars are thousands of times larger than our sun.

Astronomers have learned that many planetary systems have pairs of stars, called binary stars. In fact, scientists think most stars are in systems with more than one star!

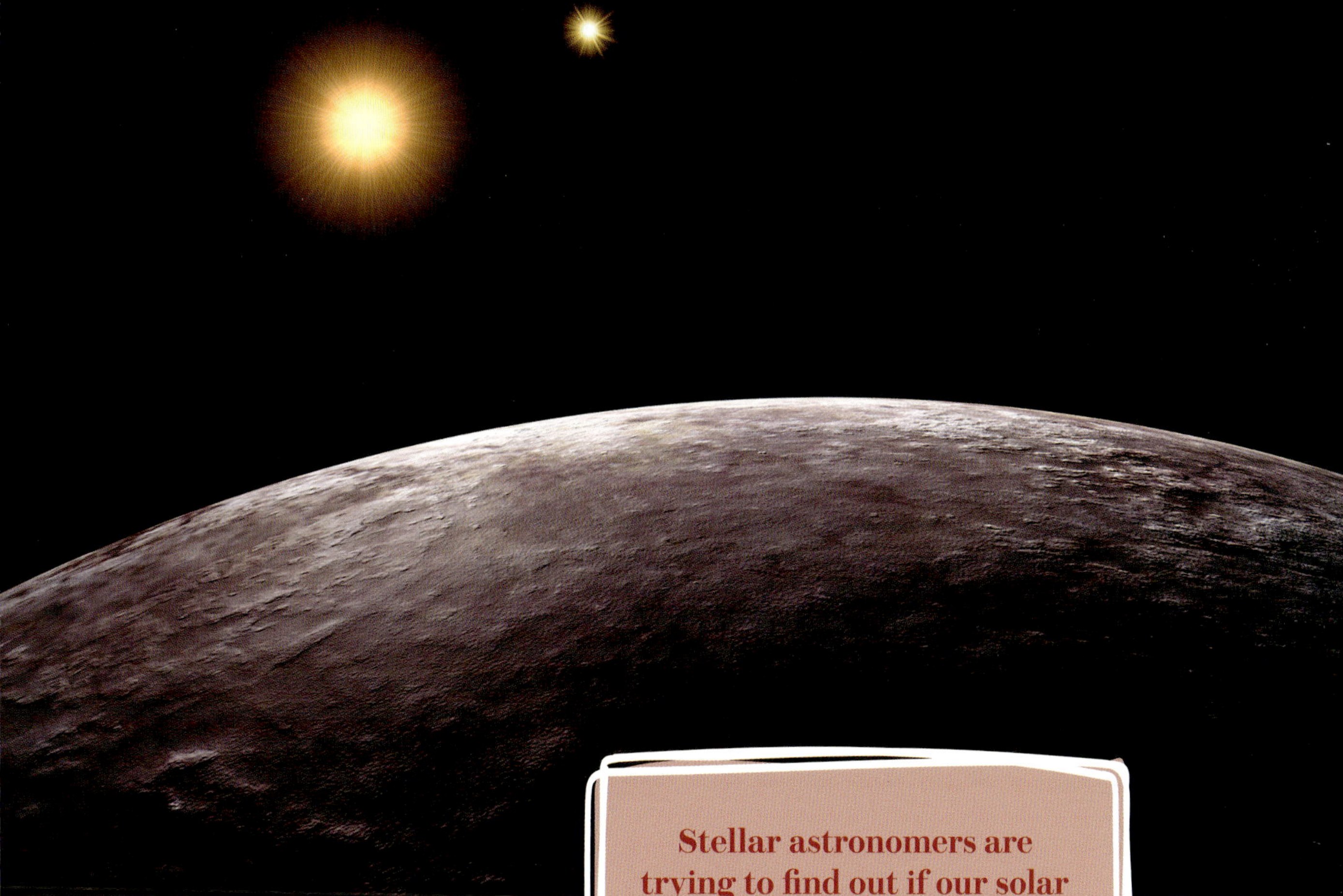

Stellar astronomers are trying to find out if our solar system was once a binary system. Can you imagine seeing two suns in the sky?

Stars that are eight times more massive than our sun or more go supernova before they die. That means they have a huge explosion. The explosion, a sudden release of energy, can shine brighter than an entire galaxy.

Astronomers can use some kinds of supernovas, called type 1a supernovas, to measure distances in space. They know how bright a supernova's explosion is. The farther away the supernova is, the dimmer it appears to be. So, stellar astronomers can use the brightness of the supernova to figure out its distance. Objects used to find distances in space are called standard candles.

artist's idea of supernova

A star's life cycle depends on its mass. Our sun is a lower-mass star and will not experience a supernova. The sun has another 6 billion years or so to burn!

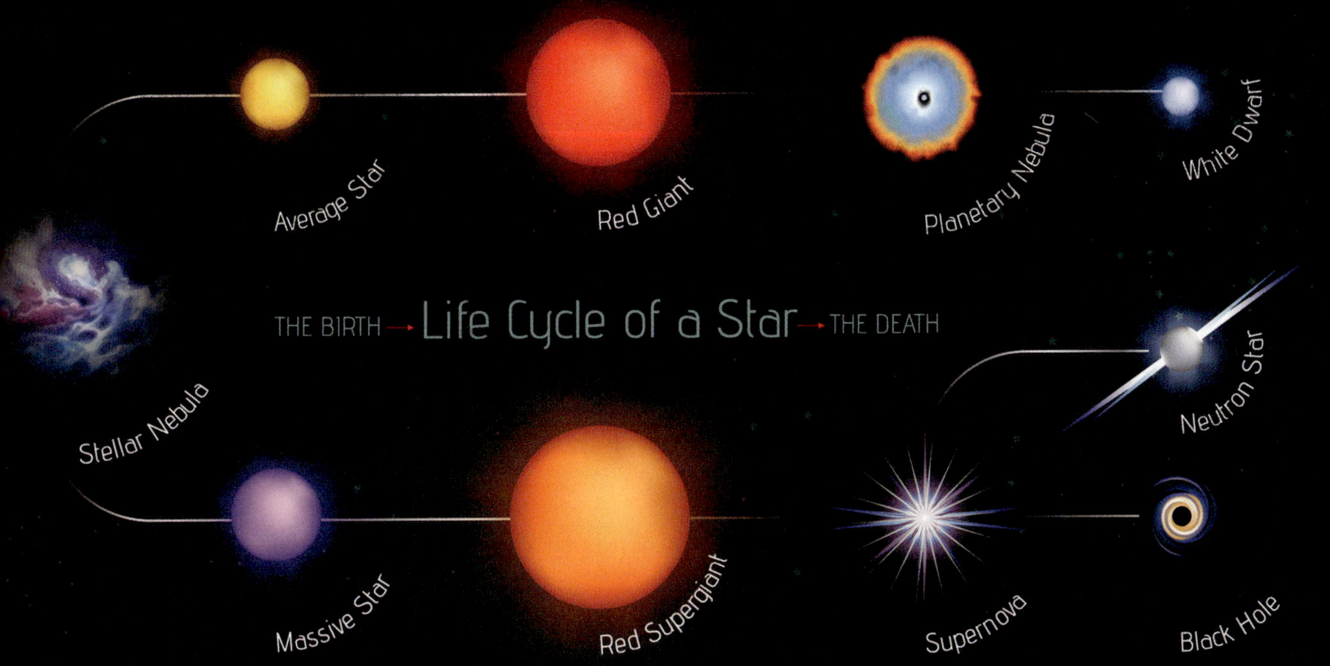

What About Our Sun?

Astronomers think our sun will become a red giant when it starts to run out of fuel. This is a large bright star with a lower surface temperature. Then, it will release a cloud of gas and dust called a planetary nebula. Next, it will shrink into a small, hot white dwarf and finally become a dark, cold black dwarf.

FINDING A BLACK HOLE

Astronomical events sometimes can't be observed, even with a telescope. That's the case with black holes. A black hole forms when a very large dying star collapses, or caves in. Its gravity becomes so strong that nothing, not even light, can escape.

Since there's no light coming from black holes, they're impossible to see. However, stellar astronomers can tell a black hole is there by tracking its effects on nearby stars and planets. For example, gas heats up as it's pulled into the black hole, releasing **X-rays**. Some kinds of telescopes can detect X-rays in space and therefore locate black holes.

FUN FACT

THE EVENT HORIZON TELESCOPE IS ACTUALLY A NUMBER OF TELESCOPES AROUND THE PLANET WORKING TOGETHER TO CREATE A TELESCOPE THE SIZE OF EARTH!

This is a black hole, called Sagittarius A*, in the center of the Milky Way galaxy. The Event Horizon Telescope, which detects radio waves, and other telescopes, including the Chandra X-ray Observatory, worked together to capture this image in 2022.

What's a Neutron Star?

Only the most massive stars leave behind black holes when they die. Large stars with less mass than these become neutron stars. After they run out of fuel, the center of the star collapses and becomes super dense, or very packed together. In fact, neutron stars are the most dense objects known.

LET'S GET STARTED

Can you imagine making a discovery about black holes, supernovas, asteroids, or anything else in this book? Consider a career in astronomy! Astronomers spend many years after high school learning as much as they can about the universe. In college, they take classes such as physics, math, computer science, and, of course, astronomy. After college, they go on to earn higher degrees.

Many astronomers become teachers and professors. Others work at NASA (National Aeronautics and Space Administration) or at a ground observatory, such as Kitt Peak National Observatory in Arizona and Cerro Tololo Inter-American Observatory in the country of Chile.

The Keck Observatory, which is made up of two optical telescopes, is located near the top of Mauna Kea, a volcano on Hawaii.

FUN FACT

BEGUN IN 1889, THE ASTRONOMICAL SOCIETY OF THE PACIFIC (ASP) IS FOR ANYONE INTERESTED IN ASTRONOMY. THE ASP IS THE LARGEST GENERAL ASTRONOMY SOCIETY IN THE WORLD, WITH MEMBERS FROM MORE THAN 70 COUNTRIES.

You don't have to wait until college or even high school to prepare for an astronomy career. You can start building your knowledge of the universe now. Study hard in your science and math classes. They'll help you understand challenging ideas in your college classes. Perfect your research and writing skills so you'll be able to tell others about your findings.

Read as much as you can about astronomy. Check your local library for the subjects that interest you most. Check NASA's website often for the latest space news. And keep looking up into the night sky. You never know what you might discover!

Science museums are another fun way to learn a lot more about astronomy.

Your First Stop

A planetarium is a place where you can see what the night sky looks like, no matter the time of day. It has a large room with a curved ceiling and many seats. A machine shines images on the ceiling and shows the stars and other objects in the night sky. Many planetariums also have telescopes.

GLOSSARY

amateur: Someone who does something without pay and not as a job.

comet: A space object made of ice and dust that has a long glowing tail when it passes close to the sun.

galaxy: A large group of stars, planets, gas, and dust that form a unit within the universe.

geology: The science that studies the history of land as recorded in rocks.

molecule: A very small piece of matter.

navigation: The science of plotting and following a path from one place to another.

optical: Having to do with the sense of sight.

physics: The study of matter, energy, force, and motion, and the relationship among them.

research: Studying to find something new.

satellite: An object that circles a planet or other space object.

space probe: An uncrewed spaceship sent to study objects in the solar system.

technology: Using science, engineering, and other industries to invent useful tools or to solve problems. Also, a machine, piece of equipment, or method created by technology.

X-ray: A powerful type of energy similar to light but invisible to the human eye.

FOR MORE INFORMATION

Books

Alméras, Gaëlle. *Super Space Weekend: Adventures in Astronomy.* Vancouver, BC, Canada: Greystone Kids, 2023.

Morrison, Marie. *20 Things You Didn't Know About Astronomy.* New York, NY: PowerKids Press, 2023.

Regas, Dean. *1,000 Facts About Space.* Washington, DC.: National Geographic Kids, 2022.

Websites

American Museum of Natural History: Journey to the Stars
www.amnh.org/explore/ology/astronomy/in-pictures-journey-to-the-stars2
Check out some amazing images of objects and events in space.

Guiness World Records: Rafał Biros
kids.guinnessworldrecords.com/records/young-achievers/rafal-biros
Learn how a 12-year-old discovered comets for NASA.

NASA: StarChild
starchild.gsfc.nasa.gov/docs/StarChild/StarChild.html
This is NASA's "learning center for young astronomers."

Publisher's note to educators and parents: Our editors have carefully reviewed these websites to ensure that they are suitable for students. Many websites change frequently, however, and we cannot guarantee that a site's future contents will continue to meet our high standards of quality and educational value. Be advised that students should be closely supervised whenever they access the internet.

INDEX